THIS
PODCAST
PLANNER

Belongs To:

DEDICATION

This Podcast Planner Log book is dedicated to all the amazing podcasters out there who love to produce podcast episodes and document their findings in the process.

You are my inspiration for producing books and I'm honored to be a part of keeping all of your Podcast notes and records organized.

This journal notebook will help you record your details about Podcasting.

Thoughtfully put together with these sections to record: Podcast Name & Episode #, Recording Date & Location, Broadcast Date, Host, Guests, Main Feature, Running Order, Music/ FX, Contest & Talking Points.

HOW TO USE THIS BOOK:

The purpose of this book is to keep all of your Podcast notes all in one place. It will help keep you organized.

This Podcast Planner Journal will allow you to accurately document every detail about your Podcast. It's a great way to chart your course through Podcasting.

Here are examples of the prompts for you to fill in and write about your experience in this book:

1. Podcast Name & Episode # - Write the Name of your Podcast and the Episode number.

2. Recording Date, Broadcast Date, & Recording Location - Log the date and location details.

3. Host - Record the Host's name.

4. Guests - For writing any Guests you may have.

5. Main Feature - Write your Main Topic for the episode.

6. Running Order - Log any specifics for Time Stamp & Segment.

7. Music/ FX - Write which Music to play.

8. Contest - Record the Sponsor, Prize and the Winner.

9. Talking Points - Blank Lined Notes for writing any other important information you want to bring up and talk about.

Enjoy!

Podcast Name _____ **Episode #** _____

Recording Date _____ Broadcast Date _____

Recording Location _____

Host(s) _____

Guest(s) _____ Fee? _____

Main Feature _____

Running Order

Time Stamp	Segment

Music / FX _____

Contest

Sponsor _____

Prize _____

Winner _____

Talking Points

_____

_____

_____

_____

Podcast Name _____ **Episode #** _____

Recording Date _____ Broadcast Date _____

Recording Location _____

Host(s) _____

Guest(s) _____ Fee? _____

Main Feature _____

Running Order

Time Stamp	Segment

Music / FX _____

Contest

Sponsor _____

Prize _____

Winner _____

Talking Points

_____

_____

_____

_____

Podcast Name _____ **Episode #** _____

Recording Date _____ Broadcast Date _____

Recording Location _____

Host(s) _____

Guest(s) _____ Fee? _____

Main Feature _____

Running Order

Time Stamp	Segment

Music / FX _____

Contest

Sponsor _____

Prize _____

Winner _____

Talking Points

_____

_____

_____

_____

Podcast Name _____ **Episode #** _____

Recording Date _____ Broadcast Date _____

Recording Location _____

Host(s) _____

Guest(s) _____ Fee? _____

Main Feature _____

Running Order

Time Stamp	Segment

Music / FX _____

Contest

Sponsor _____

Prize _____

Winner _____

Talking Points

_____

_____

_____

_____

Podcast Name _____ **Episode #** _____

Recording Date _____ Broadcast Date _____

Recording Location _____

Host(s) _____

Guest(s) _____ Fee? _____

Main Feature _____

Running Order

Time Stamp	Segment

Music / FX _____

Contest

Sponsor _____

Prize _____

Winner _____

Talking Points

_____

_____

_____

_____

Podcast Name _____ **Episode #** _____

Recording Date _____ Broadcast Date _____

Recording Location _____

Host(s) _____

Guest(s) _____ Fee? _____

Main Feature _____

Running Order

Time Stamp	Segment

Music / FX _____

Contest

Sponsor _____

Prize _____

Winner _____

Talking Points

_____

_____

_____

_____

Podcast Name _____ **Episode #** _____

Recording Date _____ Broadcast Date _____

Recording Location _____

Host(s) _____

Guest(s) _____ Fee? _____

Main Feature _____

Running Order

Time Stamp	Segment

Music / FX _____

Contest

Sponsor _____

Prize _____

Winner _____

Talking Points

_____

_____

_____

_____

Podcast Name _____ **Episode #** _____

Recording Date _____ Broadcast Date _____

Recording Location _____

Host(s) _____

Guest(s) _____ Fee? _____

Main Feature _____

Running Order

Time Stamp	Segment

Music / FX _____

Contest

Sponsor _____

Prize _____

Winner _____

Talking Points

_____

_____

_____

_____

Podcast Name _____ **Episode #** _____

Recording Date _____ Broadcast Date _____

Recording Location _____

Host(s) _____

Guest(s) _____ Fee? _____

Main Feature _____

Running Order

Time Stamp	Segment

Music / FX _____

Contest

Sponsor _____

Prize _____

Winner _____

Talking Points

_____

_____

_____

_____

Podcast Name _____ **Episode #** _____

Recording Date _____ Broadcast Date _____

Recording Location _____

Host(s) _____

Guest(s) _____ Fee? _____

Main Feature _____

Running Order

Time Stamp	Segment

Music / FX _____

Contest

Sponsor _____

Prize _____

Winner _____

Talking Points

_____

_____

_____

_____

Podcast Name _____ **Episode #** _____

Recording Date _____ Broadcast Date _____

Recording Location _____

Host(s) _____

Guest(s) _____ Fee? _____

Main Feature _____

Running Order

Time Stamp	Segment

Music / FX _____

Contest

Sponsor _____

Prize _____

Winner _____

Talking Points

_____

_____

_____

_____

Podcast Name _____ **Episode #** _____

Recording Date _____ Broadcast Date _____

Recording Location _____

Host(s) _____

Guest(s) _____ Fee? _____

Main Feature _____

Running Order

Time Stamp	Segment

Music / FX _____

Contest

Sponsor _____

Prize _____

Winner _____

Talking Points

_____

_____

_____

_____

Podcast Name _____ **Episode #** _____

Recording Date _____ Broadcast Date _____

Recording Location _____

Host(s) _____

Guest(s) _____ Fee? _____

Main Feature _____

Running Order

Time Stamp	Segment

Music / FX _____

Contest

Sponsor _____

Prize _____

Winner _____

Talking Points

_____

_____

_____

_____

Podcast Name _____ **Episode #** _____

Recording Date _____ Broadcast Date _____

Recording Location _____

Host(s) _____

Guest(s) _____ Fee? _____

Main Feature _____

Running Order

Time Stamp	Segment

Music / FX _____

Contest

Sponsor _____

Prize _____

Winner _____

Talking Points

_____

_____

_____

_____

Podcast Name _____ **Episode #** _____

Recording Date _____ Broadcast Date _____

Recording Location _____

Host(s) _____

Guest(s) _____ Fee? _____

Main Feature _____

Running Order

Time Stamp	Segment

Music / FX _____

Contest

Sponsor _____

Prize _____

Winner _____

Talking Points

_____

_____

_____

_____

Podcast Name _____ **Episode #** _____

Recording Date _____ Broadcast Date _____

Recording Location _____

Host(s) _____

Guest(s) _____ Fee? _____

Main Feature _____

Running Order

Time Stamp	Segment

Music / FX _____

Contest

Sponsor _____

Prize _____

Winner _____

Talking Points

_____

_____

_____

_____

Podcast Name _____ **Episode #** _____

Recording Date _____ Broadcast Date _____

Recording Location _____

Host(s) _____

Guest(s) _____ Fee? _____

Main Feature _____

Running Order

Time Stamp	Segment

Music / FX _____

Contest

Sponsor _____

Prize _____

Winner _____

Talking Points

_____

_____

_____

_____

Podcast Name _____ **Episode #** _____

Recording Date _____ Broadcast Date _____

Recording Location _____

Host(s) _____

Guest(s) _____ Fee? _____

Main Feature _____

Running Order

Time Stamp	Segment

Music / FX _____

Contest

Sponsor _____

Prize _____

Winner _____

Talking Points

_____

_____

_____

_____

Podcast Name _____ **Episode #** _____

Recording Date _____ Broadcast Date _____

Recording Location _____

Host(s) _____

Guest(s) _____ Fee? _____

Main Feature _____

Running Order

Time Stamp	Segment

Music / FX _____

Contest

Sponsor _____

Prize _____

Winner _____

Talking Points

_____

_____

_____

_____

Podcast Name _____ **Episode #** _____

Recording Date _____ Broadcast Date _____

Recording Location _____

Host(s) _____

Guest(s) _____ Fee? _____

Main Feature _____

Running Order

Time Stamp	Segment

Music / FX _____

Contest

Sponsor _____

Prize _____

Winner _____

Talking Points

_____

_____

_____

_____

Podcast Name _____ **Episode #** _____

Recording Date _____ Broadcast Date _____

Recording Location _____

Host(s) _____

Guest(s) _____ Fee? _____

Main Feature _____

Running Order

Time Stamp	Segment

Music / FX _____

Contest

Sponsor _____

Prize _____

Winner _____

Talking Points

_____

_____

_____

_____

Podcast Name _____ **Episode #** _____

Recording Date _____ Broadcast Date _____

Recording Location _____

Host(s) _____

Guest(s) _____ Fee? _____

Main Feature _____

Running Order

Time Stamp	Segment

Music / FX _____

Contest

Sponsor _____

Prize _____

Winner _____

Talking Points

_____

_____

_____

_____

Podcast Name _____ **Episode #** _____

Recording Date _____ Broadcast Date _____

Recording Location _____

Host(s) _____

Guest(s) _____ Fee? _____

Main Feature _____

Running Order

Time Stamp	Segment

Music / FX _____

Contest

Sponsor _____

Prize _____

Winner _____

Talking Points

_____

_____

_____

_____

Podcast Name _____ **Episode #** _____

Recording Date _____ Broadcast Date _____

Recording Location _____

Host(s) _____

Guest(s) _____ Fee? _____

Main Feature _____

Running Order

Time Stamp	Segment

Music / FX _____

Contest

Sponsor _____

Prize _____

Winner _____

Talking Points

_____

_____

_____

_____

Podcast Name _____ **Episode #** _____

Recording Date _____ Broadcast Date _____

Recording Location _____

Host(s) _____

Guest(s) _____ Fee? _____

Main Feature _____

Running Order

Time Stamp	Segment

Music / FX _____

Contest

Sponsor _____

Prize _____

Winner _____

Talking Points

_____

_____

_____

_____

Podcast Name _____ **Episode #** _____

Recording Date _____ Broadcast Date _____

Recording Location _____

Host(s) _____

Guest(s) _____ Fee? _____

Main Feature _____

Running Order

Time Stamp	Segment

Music / FX _____

Contest

Sponsor _____

Prize _____

Winner _____

Talking Points

_____

_____

_____

_____

Podcast Name _____ **Episode #** _____

Recording Date _____ Broadcast Date _____

Recording Location _____

Host(s) _____

Guest(s) _____ Fee? _____

Main Feature _____

Running Order

Time Stamp	Segment

Music / FX _____

Contest

Sponsor _____

Prize _____

Winner _____

Talking Points

_____

_____

_____

_____

Podcast Name _____ **Episode #** _____

Recording Date _____ Broadcast Date _____

Recording Location _____

Host(s) _____

Guest(s) _____ Fee? _____

Main Feature _____

Running Order

Time Stamp	Segment

Music / FX _____

Contest

Sponsor _____

Prize _____

Winner _____

Talking Points

_____

_____

_____

_____

Podcast Name _____ **Episode #** _____

Recording Date _____ Broadcast Date _____

Recording Location _____

Host(s) _____

Guest(s) _____ Fee? _____

Main Feature _____

Running Order

Time Stamp	Segment

Music / FX _____

Contest

Sponsor _____

Prize _____

Winner _____

Talking Points

_____

_____

_____

_____

Podcast Name _____ **Episode #** _____

Recording Date _____ Broadcast Date _____

Recording Location _____

Host(s) _____

Guest(s) _____ Fee? _____

Main Feature _____

Running Order

Time Stamp	Segment

Music / FX _____

Contest

Sponsor _____

Prize _____

Winner _____

Talking Points

_____

_____

_____

_____

Podcast Name _____   **Episode #** _____

Recording Date _____   Broadcast Date _____

Recording Location _____

Host(s) _____

Guest(s) _____   Fee? _____

Main Feature _____

Running Order

Time Stamp	Segment

Music / FX _____

Contest

Sponsor _____

Prize _____

Winner _____

Talking Points

_____

_____

_____

_____

Podcast Name _____ **Episode #** _____

Recording Date _____ Broadcast Date _____

Recording Location _____

Host(s) _____

Guest(s) _____ Fee? _____

Main Feature _____

Running Order

Time Stamp	Segment

Music / FX _____

Contest

Sponsor _____

Prize _____

Winner _____

Talking Points

_____

_____

_____

_____

Podcast Name _____ **Episode #** _____

Recording Date _____ Broadcast Date _____

Recording Location _____

Host(s) _____

Guest(s) _____ Fee? _____

Main Feature _____

Running Order

Time Stamp	Segment

Music / FX _____

Contest

Sponsor _____

Prize _____

Winner _____

Talking Points

_____

_____

_____

_____

Podcast Name _____ **Episode #** _____

Recording Date _____ Broadcast Date _____

Recording Location _____

Host(s) _____

Guest(s) _____ Fee? _____

Main Feature _____

Running Order

Time Stamp	Segment

Music / FX _____

Contest

Sponsor _____

Prize _____

Winner _____

Talking Points

_____

_____

_____

_____

Podcast Name _____ **Episode #** _____

Recording Date _____ Broadcast Date _____

Recording Location _____

Host(s) _____

Guest(s) _____ Fee? _____

Main Feature _____

Running Order

Time Stamp	Segment

Music / FX _____

Contest

Sponsor _____

Prize _____

Winner _____

Talking Points

_____

_____

_____

_____

Podcast Name _____ **Episode #** _____

Recording Date _____ Broadcast Date _____

Recording Location _____

Host(s) _____

Guest(s) _____ Fee? _____

Main Feature _____

Running Order

Time Stamp	Segment

Music / FX _____

Contest

Sponsor _____

Prize _____

Winner _____

Talking Points

_____

_____

_____

_____

Podcast Name _____ **Episode #** _____

Recording Date _____ Broadcast Date _____

Recording Location _____

Host(s) _____

Guest(s) _____ Fee? _____

Main Feature _____

Running Order

Time Stamp	Segment

Music / FX _____

Contest

Sponsor _____

Prize _____

Winner _____

Talking Points

_____

_____

_____

_____

Podcast Name _____ **Episode #** _____

Recording Date _____ Broadcast Date _____

Recording Location _____

Host(s) _____

Guest(s) _____ Fee? _____

Main Feature _____

Running Order

Time Stamp	Segment

Music / FX _____

Contest

Sponsor _____

Prize _____

Winner _____

Talking Points

_____

_____

_____

_____

Podcast Name _____ **Episode #** _____

Recording Date _____ Broadcast Date _____

Recording Location _____

Host(s) _____

Guest(s) _____ Fee? _____

Main Feature _____

Running Order

Time Stamp	Segment

Music / FX _____

Contest

Sponsor _____

Prize _____

Winner _____

Talking Points

_____

_____

_____

_____

Podcast Name _____ **Episode #** _____

Recording Date _____ Broadcast Date _____

Recording Location _____

Host(s) _____

Guest(s) _____ Fee? _____

Main Feature _____

Running Order

Time Stamp	Segment

Music / FX _____

Contest

Sponsor _____

Prize _____

Winner _____

Talking Points

_____

_____

_____

_____

Podcast Name _____ **Episode #** _____

Recording Date _____ Broadcast Date _____

Recording Location _____

Host(s) _____

Guest(s) _____ Fee? _____

Main Feature _____

Running Order

Time Stamp	Segment

Music / FX _____

Contest

Sponsor _____

Prize _____

Winner _____

Talking Points

_____

_____

_____

_____

Podcast Name _____ **Episode #** _____

Recording Date _____ Broadcast Date _____

Recording Location _____

Host(s) _____

Guest(s) _____ Fee? _____

Main Feature _____

Running Order

Time Stamp	Segment

Music / FX _____

Contest

Sponsor _____

Prize _____

Winner _____

Talking Points

_____

_____

_____

_____

Podcast Name _____ **Episode #** _____

Recording Date _____ Broadcast Date _____

Recording Location _____

Host(s) _____

Guest(s) _____ Fee? _____

Main Feature _____

Running Order

Time Stamp	Segment

Music / FX _____

Contest

Sponsor _____

Prize _____

Winner _____

Talking Points

_____

_____

_____

_____

Podcast Name _____ **Episode #** _____

Recording Date _____ Broadcast Date _____

Recording Location _____

Host(s) _____

Guest(s) _____ Fee? _____

Main Feature _____

Running Order

Time Stamp	Segment

Music / FX _____

Contest

Sponsor _____

Prize _____

Winner _____

Talking Points

_____

_____

_____

_____

Podcast Name _____ **Episode #** _____

Recording Date _____ Broadcast Date _____

Recording Location _____

Host(s) _____

Guest(s) _____ Fee? _____

Main Feature _____

Running Order

Time Stamp	Segment

Music / FX _____

Contest

Sponsor _____

Prize _____

Winner _____

Talking Points

_____

_____

_____

_____

Podcast Name _____ **Episode #** _____

Recording Date _____ Broadcast Date _____

Recording Location _____

Host(s) _____

Guest(s) _____ Fee? _____

Main Feature _____

Running Order

Time Stamp	Segment

Music / FX _____

Contest

Sponsor _____

Prize _____

Winner _____

Talking Points

_____

_____

_____

_____

Podcast Name _____ **Episode #** _____

Recording Date _____ Broadcast Date _____

Recording Location _____

Host(s) _____

Guest(s) _____ Fee? _____

Main Feature _____

Running Order

Time Stamp	Segment

Music / FX _____

Contest

Sponsor _____

Prize _____

Winner _____

Talking Points

_____

_____

_____

_____

Podcast Name _____ **Episode #** _____

Recording Date _____ Broadcast Date _____

Recording Location _____

Host(s) _____

Guest(s) _____ Fee? _____

Main Feature _____

Running Order

Time Stamp	Segment

Music / FX _____

Contest

Sponsor _____

Prize _____

Winner _____

Talking Points

_____

_____

_____

_____

Podcast Name _____ **Episode #** _____

Recording Date _____ Broadcast Date _____

Recording Location _____

Host(s) _____

Guest(s) _____ Fee? _____

Main Feature _____

Running Order

Time Stamp	Segment

Music / FX _____

Contest

Sponsor _____

Prize _____

Winner _____

Talking Points

_____

_____

_____

_____

Podcast Name _____ **Episode #** _____

Recording Date _____ Broadcast Date _____

Recording Location _____

Host(s) _____

Guest(s) _____ Fee? _____

Main Feature _____

Running Order

Time Stamp	Segment

Music / FX _____

Contest

Sponsor _____

Prize _____

Winner _____

Talking Points

_____

_____

_____

_____

Podcast Name _____ **Episode #** _____

Recording Date _____ Broadcast Date _____

Recording Location _____

Host(s) _____

Guest(s) _____ Fee? _____

Main Feature _____

Running Order

Time Stamp	Segment

Music / FX _____

Contest

Sponsor _____

Prize _____

Winner _____

Talking Points

_____

_____

_____

_____

Podcast Name _____ **Episode #** _____

Recording Date _____ Broadcast Date _____

Recording Location _____

Host(s) _____

Guest(s) _____ Fee? _____

Main Feature _____

Running Order

Time Stamp	Segment

Music / FX _____

Contest

Sponsor _____

Prize _____

Winner _____

Talking Points

_____

_____

_____

_____

Podcast Name _____ **Episode #** _____

Recording Date _____ Broadcast Date _____

Recording Location _____

Host(s) _____

Guest(s) _____ Fee? _____

Main Feature _____

Running Order

Time Stamp	Segment

Music / FX _____

Contest

Sponsor _____

Prize _____

Winner _____

Talking Points

_____

_____

_____

_____

Podcast Name _____ **Episode #** _____

Recording Date _____ Broadcast Date _____

Recording Location _____

Host(s) _____

Guest(s) _____ Fee? _____

Main Feature _____

Running Order

Time Stamp	Segment

Music / FX _____

Contest

Sponsor _____

Prize _____

Winner _____

Talking Points

_____

_____

_____

_____

Podcast Name _____ **Episode #** _____

Recording Date _____ Broadcast Date _____

Recording Location _____

Host(s) _____

Guest(s) _____ Fee? _____

Main Feature _____

Running Order

Time Stamp	Segment

Music / FX _____

Contest

Sponsor _____

Prize _____

Winner _____

Talking Points

_____

_____

_____

_____

Podcast Name _____ **Episode #** _____

Recording Date _____ Broadcast Date _____

Recording Location _____

Host(s) _____

Guest(s) _____ Fee? _____

Main Feature _____

Running Order

Time Stamp	Segment

Music / FX _____

Contest

Sponsor _____

Prize _____

Winner _____

Talking Points

_____

_____

_____

_____

Podcast Name _____ **Episode #** _____

Recording Date _____ Broadcast Date _____

Recording Location _____

Host(s) _____

Guest(s) _____ Fee? _____

Main Feature _____

Running Order

Time Stamp	Segment

Music / FX _____

Contest

Sponsor _____

Prize _____

Winner _____

Talking Points

_____

_____

_____

_____

Podcast Name _____ **Episode #** _____

Recording Date _____ Broadcast Date _____

Recording Location _____

Host(s) _____

Guest(s) _____ Fee? _____

Main Feature _____

Running Order

Time Stamp	Segment

Music / FX _____

Contest

Sponsor _____

Prize _____

Winner _____

Talking Points

_____

_____

_____

_____

Podcast Name _____ **Episode #** _____

Recording Date _____ Broadcast Date _____

Recording Location _____

Host(s) _____

Guest(s) _____ Fee? _____

Main Feature _____

Running Order

Time Stamp	Segment

Music / FX _____

Contest

Sponsor _____

Prize _____

Winner _____

Talking Points

_____

_____

_____

_____

Podcast Name _____ **Episode #** _____

Recording Date _____ Broadcast Date _____

Recording Location _____

Host(s) _____

Guest(s) _____ Fee? _____

Main Feature _____

Running Order

Time Stamp	Segment

Music / FX _____

Contest

Sponsor _____

Prize _____

Winner _____

Talking Points

_____

_____

_____

_____

Podcast Name _____ **Episode #** _____

Recording Date _____ Broadcast Date _____

Recording Location _____

Host(s) _____

Guest(s) _____ Fee? _____

Main Feature _____

Running Order

Time Stamp	Segment

Music / FX _____

Contest

Sponsor _____

Prize _____

Winner _____

Talking Points

_____

_____

_____

_____

Podcast Name _____   **Episode #** _____

Recording Date _____   Broadcast Date _____

Recording Location _____

Host(s) _____

Guest(s) _____   Fee? _____

Main Feature _____

Running Order

Time Stamp	Segment

Music / FX _____

Contest

Sponsor _____

Prize _____

Winner _____

Talking Points

_____

_____

_____

_____

Podcast Name _____   **Episode #** _____

Recording Date _____   Broadcast Date _____

Recording Location _____

Host(s) _____

Guest(s) _____   Fee? _____

Main Feature _____

Running Order

Time Stamp	Segment

Music / FX _____

Contest

Sponsor _____

Prize _____

Winner _____

Talking Points

_____

_____

_____

_____

Podcast Name _____ **Episode #** _____

Recording Date _____ Broadcast Date _____

Recording Location _____

Host(s) _____

Guest(s) _____ **Fee?** _____

Main Feature _____

Running Order

Time Stamp	Segment

Music / FX _____

Contest

Sponsor _____

Prize _____

Winner _____

Talking Points

_____

_____

_____

_____

Podcast Name _____ **Episode #** _____

Recording Date _____ Broadcast Date _____

Recording Location _____

Host(s) _____

Guest(s) _____ Fee? _____

Main Feature _____

Running Order

Time Stamp	Segment

Music / FX _____

Contest

Sponsor _____

Prize _____

Winner _____

Talking Points

_____

_____

_____

_____

Podcast Name _____ **Episode #** _____

Recording Date _____ Broadcast Date _____

Recording Location _____

Host(s) _____

Guest(s) _____ Fee? _____

Main Feature _____

Running Order

Time Stamp	Segment

Music / FX _____

Contest

Sponsor _____

Prize _____

Winner _____

Talking Points

_____

_____

_____

_____

Podcast Name _____ **Episode #** _____

Recording Date _____ Broadcast Date _____

Recording Location _____

Host(s) _____

Guest(s) _____ Fee? _____

Main Feature _____

Running Order

Time Stamp	Segment

Music / FX _____

Contest

Sponsor _____

Prize _____

Winner _____

Talking Points

_____

_____

_____

_____

Podcast Name _____ **Episode #** _____

Recording Date _____ Broadcast Date _____

Recording Location _____

Host(s) _____

Guest(s) _____ Fee? _____

Main Feature _____

Running Order

Time Stamp	Segment

Music / FX _____

Contest

Sponsor _____

Prize _____

Winner _____

Talking Points

_____

_____

_____

_____

Podcast Name _____ **Episode #** _____

Recording Date _____ Broadcast Date _____

Recording Location _____

Host(s) _____

Guest(s) _____ Fee? _____

Main Feature _____

Running Order

Time Stamp	Segment

Music / FX _____

Contest

Sponsor _____

Prize _____

Winner _____

Talking Points

_____

_____

_____

_____

Podcast Name _____ **Episode #** _____

Recording Date _____ Broadcast Date _____

Recording Location _____

Host(s) _____

Guest(s) _____ Fee? _____

Main Feature _____

Running Order

Time Stamp	Segment

Music / FX _____

Contest

Sponsor _____

Prize _____

Winner _____

Talking Points

_____

_____

_____

_____

Podcast Name _____ **Episode #** _____

Recording Date _____ Broadcast Date _____

Recording Location _____

Host(s) _____

Guest(s) _____ Fee? _____

Main Feature _____

Running Order

Time Stamp	Segment

Music / FX _____

Contest

Sponsor _____

Prize _____

Winner _____

Talking Points

_____

_____

_____

_____

Podcast Name _____ **Episode #** _____

Recording Date _____ Broadcast Date _____

Recording Location _____

Host(s) _____

Guest(s) _____ Fee? _____

Main Feature _____

Running Order

Time Stamp	Segment

Music / FX _____

Contest

Sponsor _____

Prize _____

Winner _____

Talking Points

_____

_____

_____

_____

Podcast Name _____ **Episode #** _____

Recording Date _____ Broadcast Date _____

Recording Location _____

Host(s) _____

Guest(s) _____ Fee? _____

Main Feature _____

Running Order

Time Stamp	Segment

Music / FX _____

Contest

Sponsor _____

Prize _____

Winner _____

Talking Points

_____

_____

_____

_____

Podcast Name _____ **Episode #** _____

Recording Date _____ Broadcast Date _____

Recording Location _____

Host(s) _____

Guest(s) _____ Fee? _____

Main Feature _____

Running Order

Time Stamp	Segment

Music / FX _____

Contest

Sponsor _____

Prize _____

Winner _____

Talking Points

_____

_____

_____

_____

Podcast Name _____ **Episode #** _____

Recording Date _____ Broadcast Date _____

Recording Location _____

Host(s) _____

Guest(s) _____ Fee? _____

Main Feature _____

Running Order

Time Stamp	Segment

Music / FX _____

Contest

Sponsor _____

Prize _____

Winner _____

Talking Points

_____

_____

_____

_____

Podcast Name _____   **Episode #** _____

Recording Date _____   Broadcast Date _____

Recording Location _____

Host(s) _____

Guest(s) _____   Fee? _____

Main Feature _____

Running Order

Time Stamp	Segment

Music / FX _____

Contest

Sponsor _____

Prize _____

Winner _____

Talking Points

_____

_____

_____

_____

Podcast Name _____ **Episode #** _____

Recording Date _____ Broadcast Date _____

Recording Location _____

Host(s) _____

Guest(s) _____ Fee? _____

Main Feature _____

Running Order

Time Stamp	Segment

Music / FX _____

Contest

Sponsor _____

Prize _____

Winner _____

Talking Points

_____

_____

_____

_____

Podcast Name _____ **Episode #** _____

Recording Date _____ Broadcast Date _____

Recording Location _____

Host(s) _____

Guest(s) _____ Fee? _____

Main Feature _____

Running Order

Time Stamp	Segment

Music / FX _____

Contest

Sponsor _____

Prize _____

Winner _____

Talking Points

_____

_____

_____

_____

Podcast Name _____ **Episode #** _____

Recording Date _____ Broadcast Date _____

Recording Location _____

Host(s) _____

Guest(s) _____ Fee? _____

Main Feature _____

Running Order

Time Stamp	Segment

Music / FX _____

Contest

Sponsor _____

Prize _____

Winner _____

Talking Points

_____

_____

_____

_____

Podcast Name _____ **Episode #** _____

Recording Date _____ Broadcast Date _____

Recording Location _____

Host(s) _____

Guest(s) _____ Fee? _____

Main Feature _____

Running Order

Time Stamp	Segment

Music / FX _____

Contest

Sponsor _____

Prize _____

Winner _____

Talking Points

_____

_____

_____

_____

Podcast Name _____ **Episode #** _____

Recording Date _____ Broadcast Date _____

Recording Location _____

Host(s) _____

Guest(s) _____ Fee? _____

Main Feature _____

Running Order

Time Stamp	Segment

Music / FX _____

Contest

Sponsor _____

Prize _____

Winner _____

Talking Points

_____

_____

_____

_____

Podcast Name _____ **Episode #** _____

Recording Date _____ Broadcast Date _____

Recording Location _____

Host(s) _____

Guest(s) _____ Fee? _____

Main Feature _____

Running Order

Time Stamp	Segment

Music / FX _____

Contest

Sponsor _____

Prize _____

Winner _____

Talking Points

_____

_____

_____

_____

Podcast Name _____ **Episode #** _____

Recording Date _____ Broadcast Date _____

Recording Location _____

Host(s) _____

Guest(s) _____ Fee? _____

Main Feature _____

Running Order

Time Stamp	Segment

Music / FX _____

Contest

Sponsor _____

Prize _____

Winner _____

Talking Points

_____

_____

_____

_____

Podcast Name _____ **Episode #** _____

Recording Date _____ Broadcast Date _____

Recording Location _____

Host(s) _____

Guest(s) _____ Fee? _____

Main Feature _____

Running Order

Time Stamp	Segment

Music / FX _____

Contest

Sponsor _____

Prize _____

Winner _____

Talking Points

_____

_____

_____

_____

Podcast Name _____ **Episode #** _____

Recording Date _____ Broadcast Date _____

Recording Location _____

Host(s) _____

Guest(s) _____ Fee? _____

Main Feature _____

Running Order

Time Stamp	Segment

Music / FX _____

Contest

Sponsor _____

Prize _____

Winner _____

Talking Points

_____

_____

_____

_____

Podcast Name _____   **Episode #** _____

Recording Date _____   Broadcast Date _____

Recording Location _____

Host(s) _____

Guest(s) _____   Fee? _____

Main Feature _____

Running Order

Time Stamp	Segment

Music / FX _____

Contest

Sponsor _____

Prize _____

Winner _____

Talking Points

_____

_____

_____

_____

Podcast Name _____ **Episode #** _____

Recording Date _____ Broadcast Date _____

Recording Location _____

Host(s) _____

Guest(s) _____ Fee? _____

Main Feature _____

Running Order

Time Stamp	Segment

Music / FX _____

Contest

Sponsor _____

Prize _____

Winner _____

Talking Points

_____

_____

_____

_____

Podcast Name _____ **Episode #** _____

Recording Date _____ Broadcast Date _____

Recording Location _____

Host(s) _____

Guest(s) _____ Fee? _____

Main Feature _____

Running Order

Time Stamp	Segment

Music / FX _____

Contest

Sponsor _____

Prize _____

Winner _____

Talking Points

_____

_____

_____

_____

Podcast Name _____ **Episode #** _____

Recording Date _____ Broadcast Date _____

Recording Location _____

Host(s) _____

Guest(s) _____ Fee? _____

Main Feature _____

Running Order

Time Stamp	Segment

Music / FX _____

Contest

Sponsor _____

Prize _____

Winner _____

Talking Points

_____

_____

_____

_____

Podcast Name _____ **Episode #** _____

Recording Date _____ Broadcast Date _____

Recording Location _____

Host(s) _____

Guest(s) _____ Fee? _____

Main Feature _____

Running Order

Time Stamp	Segment

Music / FX _____

Contest

Sponsor _____

Prize _____

Winner _____

Talking Points

_____

_____

_____

_____

Podcast Name _____ **Episode #** _____

Recording Date _____ Broadcast Date _____

Recording Location _____

Host(s) _____

Guest(s) _____ Fee? _____

Main Feature _____

Running Order

Time Stamp	Segment

Music / FX _____

Contest

Sponsor _____

Prize _____

Winner _____

Talking Points

_____

_____

_____

_____

Podcast Name _____ **Episode #** _____

Recording Date _____ Broadcast Date _____

Recording Location _____

Host(s) _____

Guest(s) _____ Fee? _____

Main Feature _____

Running Order

Time Stamp	Segment

Music / FX _____

Contest

Sponsor _____

Prize _____

Winner _____

Talking Points

_____

_____

_____

_____

Podcast Name _____ **Episode #** _____

Recording Date _____ Broadcast Date _____

Recording Location _____

Host(s) _____

Guest(s) _____ Fee? _____

Main Feature _____

Running Order

Time Stamp	Segment

Music / FX _____

Contest

Sponsor _____

Prize _____

Winner _____

Talking Points

_____

_____

_____

_____

Podcast Name _____ **Episode #** _____

Recording Date _____ Broadcast Date _____

Recording Location _____

Host(s) _____

Guest(s) _____ Fee? _____

Main Feature _____

Running Order

Time Stamp	Segment

Music / FX _____

Contest

Sponsor _____

Prize _____

Winner _____

Talking Points

_____

_____

_____

_____

Podcast Name _____ **Episode #** _____

Recording Date _____ Broadcast Date _____

Recording Location _____

Host(s) _____

Guest(s) _____ Fee? _____

Main Feature _____

Running Order

Time Stamp	Segment

Music / FX _____

Contest

Sponsor _____

Prize _____

Winner _____

Talking Points

_____

_____

_____

_____

Podcast Name _____ **Episode #** _____

Recording Date _____ Broadcast Date _____

Recording Location _____

Host(s) _____

Guest(s) _____ Fee? _____

Main Feature _____

Running Order

Time Stamp	Segment

Music / FX _____

Contest

Sponsor _____

Prize _____

Winner _____

Talking Points

_____
_____
_____
_____

Podcast Name _____ **Episode #** _____

Recording Date _____ Broadcast Date _____

Recording Location _____

Host(s) _____

Guest(s) _____ Fee? _____

Main Feature _____

Running Order

Time Stamp	Segment

Music / FX _____

Contest

Sponsor _____

Prize _____

Winner _____

Talking Points

_____

_____

_____

_____

Podcast Name _____ **Episode #** _____

Recording Date _____ Broadcast Date _____

Recording Location _____

Host(s) _____

Guest(s) _____ Fee? _____

Main Feature _____

Running Order

Time Stamp	Segment

Music / FX _____

Contest

Sponsor _____

Prize _____

Winner _____

Talking Points

_____

_____

_____

_____

Podcast Name _____ **Episode #** _____

Recording Date _____ Broadcast Date _____

Recording Location _____

Host(s) _____

Guest(s) _____ Fee? _____

Main Feature _____

Running Order

Time Stamp	Segment

Music / FX _____

Contest

Sponsor _____

Prize _____

Winner _____

Talking Points

_____

_____

_____

_____

Podcast Name _____ **Episode #** _____

Recording Date _____ Broadcast Date _____

Recording Location _____

Host(s) _____

Guest(s) _____ Fee? _____

Main Feature _____

Running Order

Time Stamp	Segment

Music / FX _____

Contest

Sponsor _____

Prize _____

Winner _____

Talking Points

_____

_____

_____

_____

Podcast Name _____ **Episode #** _____

Recording Date _____ Broadcast Date _____

Recording Location _____

Host(s) _____

Guest(s) _____ Fee? _____

Main Feature _____

Running Order

Time Stamp	Segment

Music / FX _____

Contest

Sponsor _____

Prize _____

Winner _____

Talking Points

_____

_____

_____

_____

Podcast Name _____ **Episode #** _____

Recording Date _____ Broadcast Date _____

Recording Location _____

Host(s) _____

Guest(s) _____ Fee? _____

Main Feature _____

Running Order

Time Stamp	Segment

Music / FX _____

Contest

Sponsor _____

Prize _____

Winner _____

Talking Points

_____

_____

_____

_____

Podcast Name _____ **Episode #** _____

Recording Date _____ Broadcast Date _____

Recording Location _____

Host(s) _____

Guest(s) _____ Fee? _____

Main Feature _____

Running Order

Time Stamp	Segment

Music / FX _____

Contest

Sponsor _____

Prize _____

Winner _____

Talking Points

_____

_____

_____

_____

Podcast Name _____ **Episode #** _____

Recording Date _____ Broadcast Date _____

Recording Location _____

Host(s) _____

Guest(s) _____ Fee? _____

Main Feature _____

Running Order

Time Stamp	Segment

Music / FX _____

Contest

Sponsor _____

Prize _____

Winner _____

Talking Points

_____

_____

_____

_____

Podcast Name _____ **Episode #** _____

Recording Date _____ Broadcast Date _____

Recording Location _____

Host(s) _____

Guest(s) _____ Fee? _____

Main Feature _____

Running Order

Time Stamp	Segment

Music / FX _____

Contest

Sponsor _____

Prize _____

Winner _____

Talking Points

_____

_____

_____

_____

Podcast Name _____   **Episode #** _____

Recording Date _____   Broadcast Date _____

Recording Location _____

Host(s) _____

Guest(s) _____   Fee? _____

Main Feature _____

Running Order

Time Stamp	Segment

Music / FX _____

Contest

Sponsor _____

Prize _____

Winner _____

Talking Points

_____

_____

_____

_____

Podcast Name _____   **Episode #** _____

Recording Date _____   Broadcast Date _____

Recording Location _____

Host(s) _____

Guest(s) _____   Fee? _____

Main Feature _____

Running Order

Time Stamp	Segment

Music / FX _____

Contest

Sponsor _____

Prize _____

Winner _____

Talking Points

_____

_____

_____

_____

www.ingramcontent.com/pod-product-compliance
Lightning Source LLC
Chambersburg PA
CBHW080600030426
42336CB00019B/3269